WORDS FROM THE FOUNDER

Akshansh Mishra,
Founder and Project
Scientific Officer, Stir
Research Technologies

It is an immense joy for me to launch a first magazine which purely going to focus on Friction Stir Welding Research and Innovations. This magazine will be published quarterly.

Each issue of the magazine will focus on a new topic in Friction Stir Welding research.

The main objective of this magazine will be to enhance the knowledge and skills of young researchers in Friction Stir Welding technology. Friction Stir Welding Technology has opened a gateway to join non weldable alloys. This first issue is going to focus on the artificial intelligence techniques like Artificial Neural Network application in Friction Stir Welding research.

Readers are free to convey their message on akshansh.frictionwelding@gmail.com.

Any collaborations are welcome further.

How does our brain process any information?

Katyayani Jaiswal
Department of Computer Science, Indian Institute of Technology, Ropar

The human brain is a complicated, creative information-processing system. As technology advanced from primitive to modern, the metaphors used to describe the brain also advanced. Initially, it was compared to a wax tablet, then to a sheet of papyrus, then to a book, and most recently, to a computer. As you learn about the brain, keep in mind that the usefulness of these metaphors is limited and can lead to erroneous conclusions.

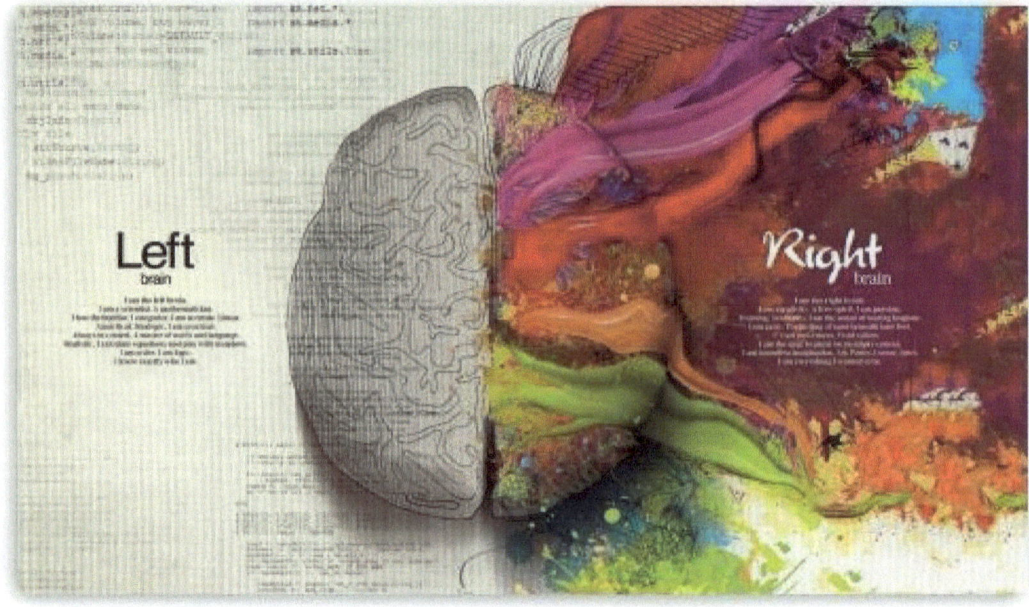

Information processing starts with input from the sensory organs, which transform physical stimuli such as touch, heat, sound waves, or photons of light into electrochemical signals. The sensory information is repeatedly transformed by the algorithms of the brain in both bottom-up and top-down processing. For example, when looking at a picture of a black box on a white background, bottom-up processing puts together very simple information such as color, orientation, and where the borders of the object are - where the color changes significantly over a short space - to decide that you are seeing a box. Top-down processing uses the decisions made at some steps of the bottom-up process to speed up your recognition of the box. Top-down processing in this example might help you identify the object as a black box rather than a box-shaped hole in the white background.

Once information is processed to a degree, an attention filter decides how important the signal is and which cognitive processes it should be made available to. For example, although your brain processes every blade of grass when you look down at your shoes, a healthy attention filter prevents you from noticing them individually. In contrast, you might pick out your name, even when spoken in a noisy room. There are many stages of processing, and the results of processing are modulated by attention repeatedly.

In order for the brain to process information, it must first be stored. There are multiple types of memory, including sensory, working, and long-term. First, information is encoded. There are types of encoding specific to each type of sensory stimuli. For example, verbal input can be encoded structurally, referring to what the printed word looks like, phonemically, referring to what the word sounds like, or semantically, referring to what the word means. Once information is stored, it must be maintained. Some animal studies suggest that working memory, which stores information for roughly 20 seconds, is maintained by an electrical signal looping through a particular series of neurons for a short period of time. Information in long-term memory is hypothesized to be maintained in the structure of certain types of proteins.

There are numerous models of how the knowledge is organized in the brain, some based on the way human subjects retrieve memories, others based on computer science, and others based on neurophysiology. The semantic network model states that there are nodes representing concepts, and that the nodes are linked based on their relatedness. For example, in a semantic network, "chair" might be linked to "table," which can be linked to "wooden," and so forth. The connectionist model states that a piece of knowledge is represented merely by a pattern of neuronal activation rather than by meaning. There is not yet a universally accepted knowledge organization model, because each has strengths and weaknesses.

2

Once stored, memories eventually must be retrieved from storage. Remembering past events is not like watching a recorded video. It is, rather, a process of reconstructing what may have happened based on the details the brain chose to store and was able to recall. Recall is triggered by a retrieval cue, an environmental stimulus that prompts the brain to retrieve the memory. Evidence shows that the better the retrieval cue, the higher the chance of recalling the memory. It is important to note that the retrieval cue can also make a person reconstruct a memory improperly. Memory distortions can be produced in various ways, including varying the wording of a question. For example, merely asking someone whether a red car had left the scene of a hit-and-run can make the person recall having seen a red car during later questioning, even if there was never a red car.

Information processing in the brain is the topic of a large, ongoing body of research. Although some people are fascinated by the brain on its own merits, a growing number are looking to psychology in order to better their own study skills and cognitive performance.

How Artificial Neural Network works?

Katyayani Jaiswal
Department of Computer Science, Indian Institute of Technology, Ropar

Artificial Neural Network (ANN) can be considered as a mathematical model of a human brain. This elemental inspired method marks the next generation advancement in the computing field. The composition of Artificial Neural Network (ANN) consists of a large number of simple processing elements or basic units called neurons. Each neuron applies an activation function to its net input to determine its output signal. Every neuron is connected to other neurons by means of directed communication links, each with an associated weight [1]. Each neuron has an internal state called its activation level, which is a function of the inputs it has received. This can be compared with a bottle with a liquid. If we have a bottle and if we fill in the bottle with a liquid, and if we have an alarm to caution us when the level of the liquid is up to the neck of the bottle, then activation level also does the same thing as that of the alarming signal we receive. As and when the neuron receives the signal, it gets added up and when the cumulative signal reaches the activation level the neuron sends an output. Till then it keeps receiving the input. So activation level can be considered as a threshold value for us to understand.

The technique is particularly suited to problems that involve the manipulation of multiple parameters and non-linear interpolation, and as a consequence are therefore not easily amenable to conventional theoretical and mathematical approaches. Neural networks have therefore seen growing application in materials property (mechanical and physical) determination, particularly the more difficult to analyse complex multiphase and composite materials, which are growing in popularity [2].

How do Artificial Neural Network Works?

This is a large and complex topic because there are many different types of artificial neural network models. The most common model, which has become the foundation for most of the others, is the 3-layer fully-connected back propagation (BP) model:

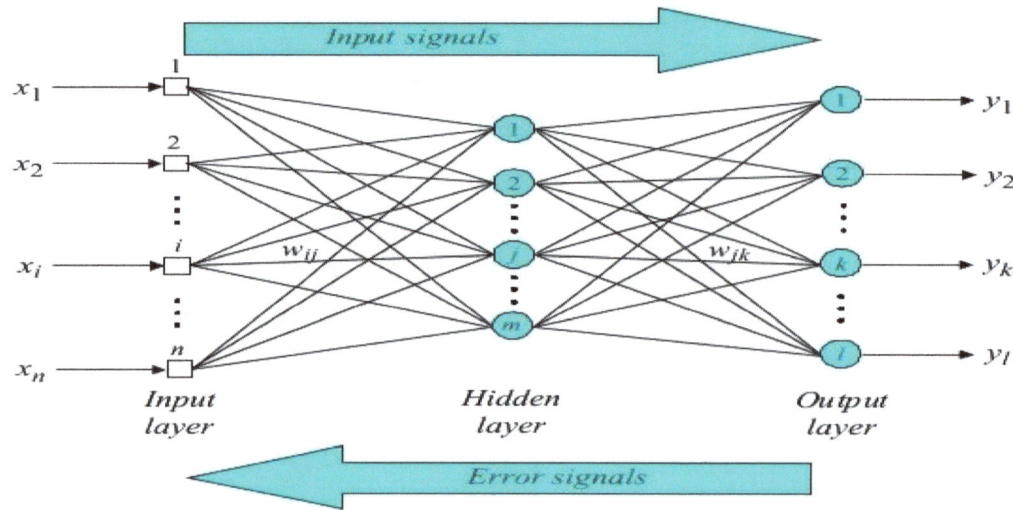

Network Design

The basic idea is that you have three layers of "nodes." The "nodes" are intended to be analogous to neurons in a neural network of the brain, but the similarity is only metaphorical (real neurons don't work this way, but the analogy is not unreasonable). The nodes have values of 0.0 to 1.0, where 0 represents fully inactive "off" and 1 represents fully active "on" with many values in between. The three layers are an input layer, an output layer, and a "hidden" layer in the middle (hidden means neither input nor output, so not exposed to the outside world). The nodes are linked by connections which have a "weight" ("w" in the figure) that are analogous to synapses in the brain. Signal values propagate from the inputs, through the connection weights to the hidden nodes, and then onward through more connection weights to the output nodes. The number of the neurons at the first and the last layer are equal to the inputs and outputs of the ANN. The user determines the number of neurons at the intermediate layer (hidden layer) with trial and error. In most of the BP applications, each neuron is connected to all the neurons of the following layer.

In the beginning, the network will of course get the wrong answer because it knows nothing. This is where the "training" and "back propagation" comes in. The error values are propagated backward through the network using some complicated math that tells the algorithm how to modify each connection weight so that the network will get closer to the correct answer next time.

Artificial neural networks are computational models which work similar to the functioning of a human nervous system. There are several kinds of artificial neural networks. These type of networks are implemented based on the mathematical operations and a set of parameters required to determine the output. Let's look at some of the neural networks:

Feedforward Neural Network – Artificial Neuron:

This neural network is one of the simplest form of ANN, where the data or the input travels in one direction. The data passes through the input nodes and exit on the output nodes. This neural network may or may not have the hidden layers. In simple words, it has a front propagated wave and no back propagation by using a classifying activation function usually.

Below is a Single layer feed forward network. Here, the sum of the products of inputs and weights are calculated and fed to the output. The output is considered if it is above a certain value i.e threshold(usually 0) and the neuron fires with an activated output (usually 1) and if it does not fire, the deactivated value is emitted (usually -1).

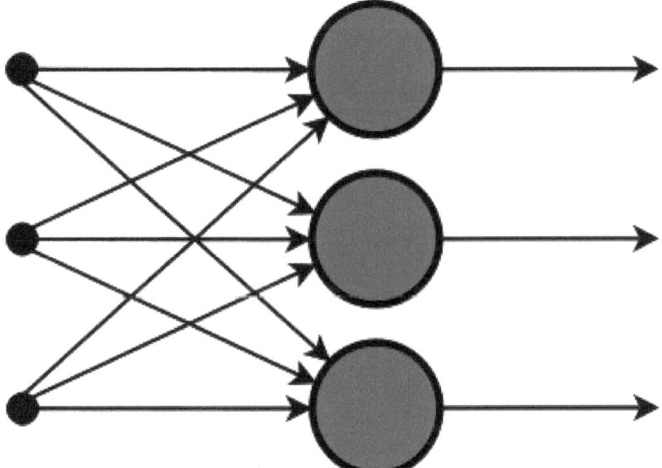

Application of Feed forward neural networks are found in computer vision and speech recognition where classifying the target classes are complicated. These kind of Neural Networks are responsive to noisy data and easy to maintain.

Radial basis function Neural Network:

Radial basic functions consider the distance of a point with respect to the center. RBF functions have two layers, first where the features are combined with the Radial Basis Function in the inner layer and then the output of these features are taken into consideration while computing the same output in the next time-step which is basically a memory.

Below is a diagram which represents the distance calculating from the center to a point in the plane similar to a radius of the circle. Here, the distance measure used in euclidean, other distance measures can also be used. The model depends on the maximum reach or the radius of the circle in classifying the points into different categories. If the point is in or around the radius, the likelihood of the new point begin classified into that class is high. There can be a transition while changing from one region to another and this can be controlled by the beta function.

This neural network has been applied in Power Restoration Systems. Power systems have increased in size and complexity. Both factors increase the risk of major power outages. After a blackout, power needs to be restored as quickly and reliably as possible. This paper how RBFnn has been implemented in this domain.

Power restoration usually proceeds in the following order:

- First priority is to restore power to essential customers in the communities. These customers provide health care and safety services to all and restoring power to them first enables them to help many others. Essential customers include health care facilities, school boards, critical municipal infrastructure, and police and fire services.
- Then focus on major power lines and substations that serve larger numbers of customers
- Give higher priority to repairs that will get the largest number of customers back in service as quickly as possible
- Then restore power to smaller neighborhoods and individual homes and businesses

The diagram below shows the typical order of power restoration system.

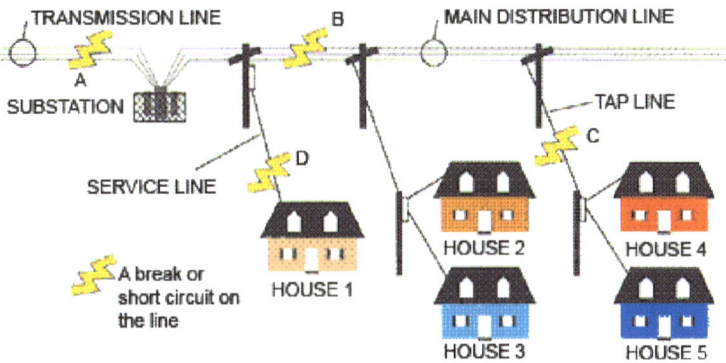

Referring to the diagram, first priority goes to fixing the problem at point A, on the transmission line. With this line out, none of the houses can have power restored. Next, fixing the problem at B on the main distribution line running out of the substation. Houses 2, 3, 4 and 5 are affected by this problem. Next, fixing the line at C, affecting houses 4 and 5. Finally, we would fix the service line at D to house 1.

Kohonen Self Organizing Neural Network:

The objective of a Kohonen map is to input vectors of arbitrary dimension to discrete map comprised of neurons. The map needs to me trained to create its own organization of the training data. It comprises of either one or two dimensions. When training the map the location of the neuron remains constant but the weights differ depending on the value. This self organization process has different parts, in the first phase every neuron value is initialized with a small weight and the input vector. In the second phase, the neuron closest to the point is the 'winning neuron' and the neurons connected to the winning neuron will also move towards the point like in the graphic below. The distance between the point and the neurons is calculated by the euclidean distance, the neuron with the least distance wins. Through the iterations, all the points are clustered and each neuron represents each kind of cluster. This is the gist behind the organization of Kohonen Neural Network.

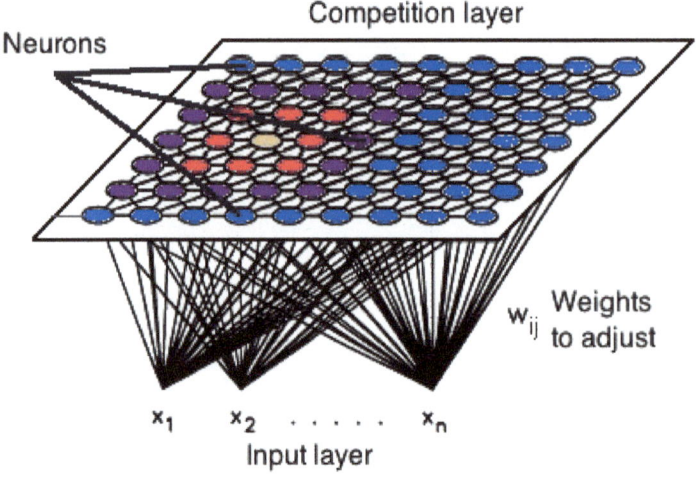

Kohonen Neural Network is used to recognize patterns in the data. Its application can be found in medical analysis to cluster data into different categories. Kohonen map was able to classify patients having glomerular or tubular with an high accuracy. Here is a detailed explanation of how it is categorized mathematically using the euclidean distance algorithm. Below is an image displaying a comparison between a healthy and a diseased glomerular.

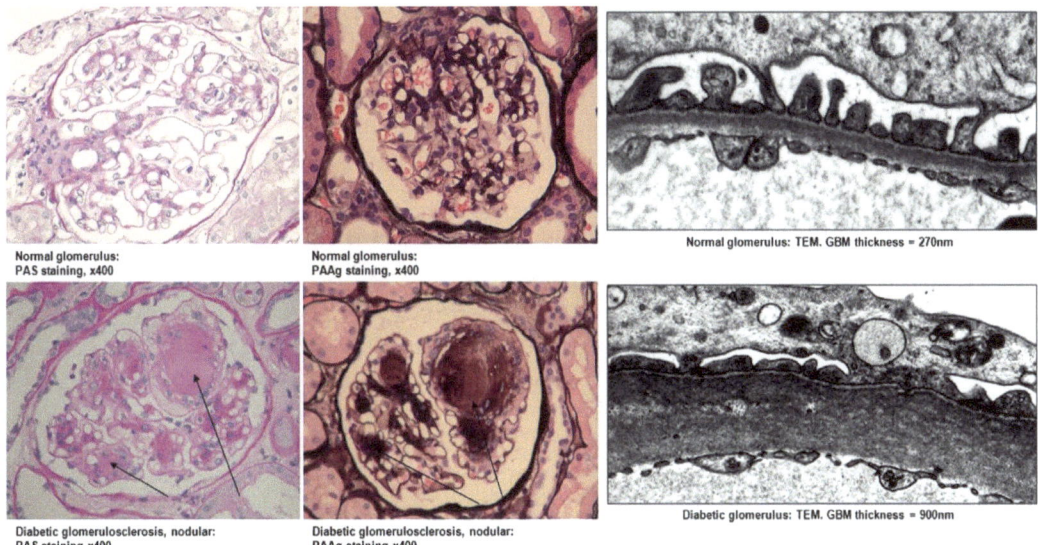

Normal glomerulus:
PAS staining, x400

Normal glomerulus:
PAAg staining, x400

Normal glomerulus: TEM. GBM thickness = 270nm

Diabetic glomerulosclerosis, nodular:
PAS staining x400

Diabetic glomerulosclerosis, nodular:
PAAg staining x400

Diabetic glomerulus: TEM. GBM thickness = 900nm

Recurrent Neural Network(RNN) – Long Short Term Memory:

The Recurrent Neural Network works on the principle of saving the output of a layer and feeding this back to the input to help in predicting the outcome of the layer.

Here, the first layer is formed similar to the feed forward neural network with the product of the sum of the weights and the features. The recurrent neural network process starts once this is computed, this means that from one time step to the next each neuron will remember some information it had in the previous time-step. This makes each neuron act like a memory cell in performing computations. In this process, we need to let the neural network to work on the front propagation and remember what information it needs for later use. Here, if the prediction is wrong we use the learning rate or error correction to make small changes so that it will gradually work towards making the right prediction during the back propagation. This is how a basic Recurrent Neural Network looks like,

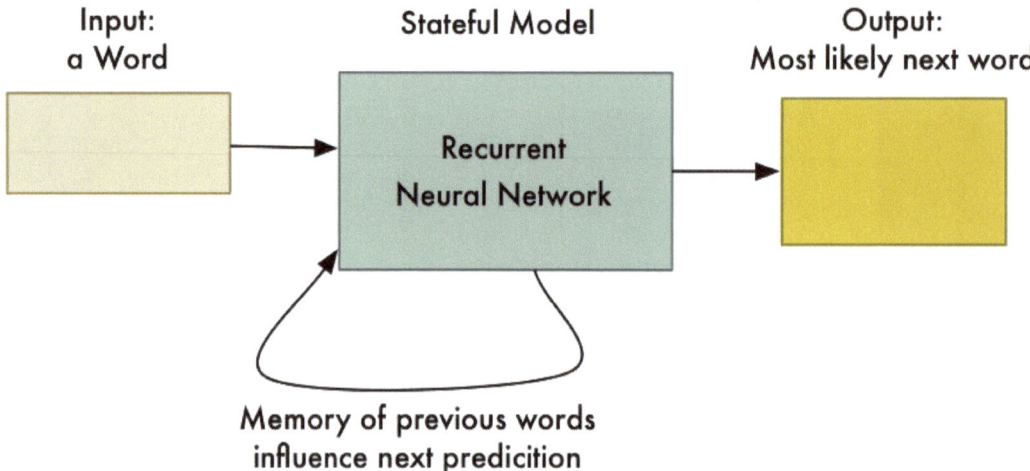

Output so far:

Machine

Convolutional Neural Network:

Convolutional neural networks are similar to feed forward neural networks , where the neurons have learn-able weights and biases. Its application have been in signal and image processing which takes over OpenCV in field of computer vision.

Below is a representation of a ConvNet, in this neural network, the input features are taken in batch wise like a filter. This will help the network to remember the images in parts and can compute the operations. These computations involve conversion of the image from RGB or HSI scale to Gray-scale. Once we have this, the changes in the pixel value will help detecting the edges and images can be classified into different categories.

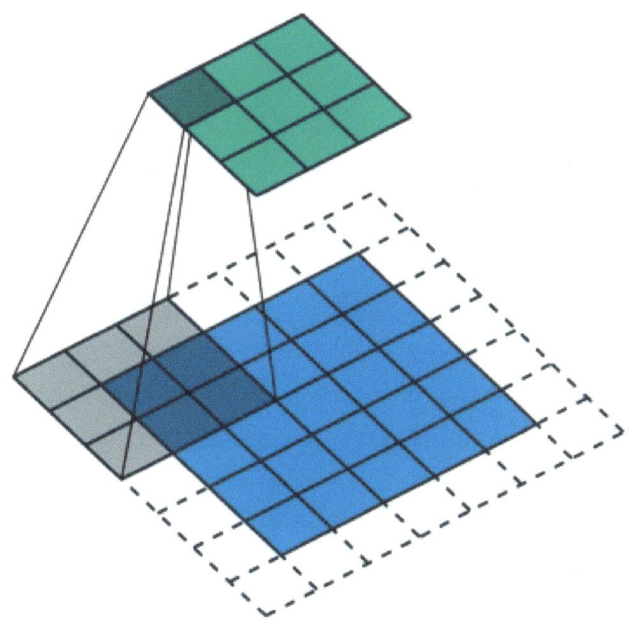

ConvNet are applied in techniques like signal processing and image classification techniques. Computer vision techniques are dominated by convolutional neural networks because of their accuracy in image classification. The technique of image analysis and recognition, where the agriculture and weather features are extracted from the open source satellites like LSAT to predict the future growth and yield of a particular land are being implemented.

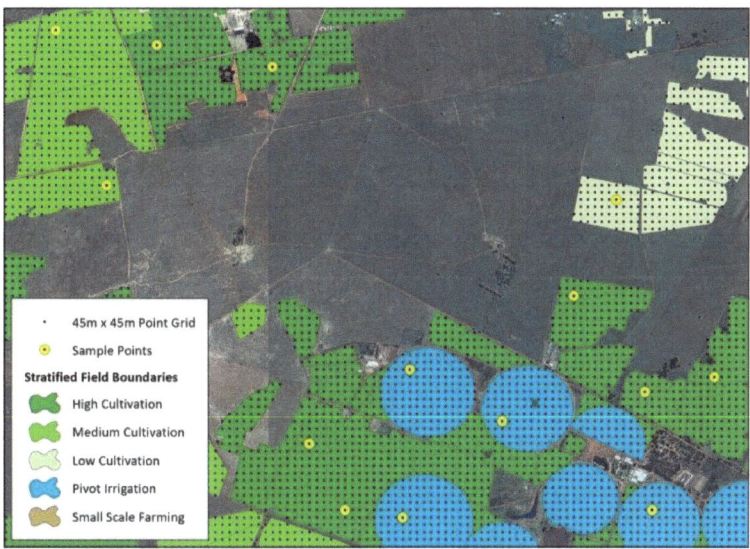

Modular Neural Network:

Modular Neural Networks have a collection of different networks working independently and contributing towards the output. Each neural network has a set of inputs which are unique compared to other networks constructing and performing sub-tasks. These networks do not interact or signal each other in accomplishing the tasks. The advantage of a modular neural network is that it breakdowns a large computational process into smaller components decreasing the complexity. This breakdown will help in decreasing the number of connections and negates the interaction of these network with each other, which in turn will increase the computation speed. However, the processing time will depend on the number of neurons and their involvement in computing the results.

Below is a visual representation,

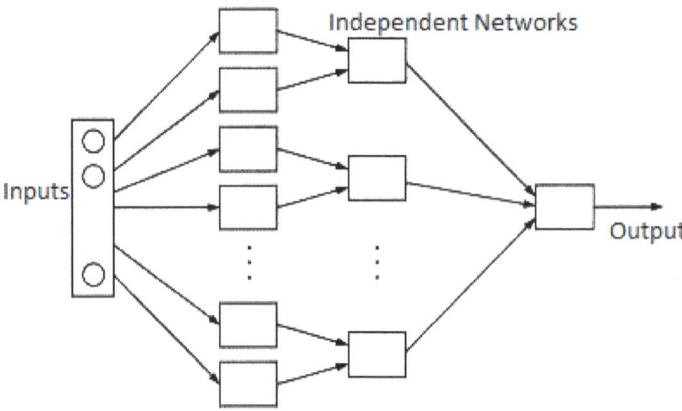

Source: Analytics India Mag

Use of Artificial Neural Network in Friction Stir Welding Research

Akshansh Mishra
Founder and Project Scientific Officer, Stir Research Technologies

Artificial Neural Network (ANN) is a brain modelling technique by providing a new approach to computing. It introduces a less technical way to develop machine solutions. This article discusses the use of Artificial Neural Network (ANN) concept in Friction Stir Welding research, for example it is used in the investigation of tool parameters, for the evaluation of feedback forces which is provided by Friction Stir Welding process. Previous research also shows that ANN finds application in developing the correlation between the Friction Stir Welding parameters of the light alloy plates and mechanical properties. This method was also used for predicting average grain size in Friction Stir Welding processes.

Friction Stir is a solid state joining process developed by The Welding Institute (TWI) in the UK in 1991. This method is used for joining the alloys of aluminium, magnesium, copper, titanium and as well as steel plates [3-8]. Artificial Neural Network (ANN) plays an important role in Friction Stir Welding (FSW) Research. It is basically used to develop the applications of Friction Stir Welding (FSW) and reduce the cost of experiments. Tansel et al [9] represented the characteristics of Friction Stir Welding process by using Artificial Neural Network (ANN). Dehabadi et al [10] predicted the Vickers micro hardness of AA6061 Friction Stir Welded sheets by using Artificial Neural Network (ANN). Shojaeefard et al [11] performed Artificial Neural Network (ANN) analysis to model the correlation between the tool parameters (pin and shoulder diameter) and heat-affected zone, thermal, and strain value in the weld zone. Fratini et al [12] linked Artificial Neural Network to a finite element model (FEM) and predicted the average grain size values of butt, lap and T friction stir welded joints. Jayaraman et al [13] by Artificial Neural Network modelling predicted the tensile strength of A356 alloy which is a high strength Aluminium-Silicon cast alloy used in food, chemical, marine, electrical and automotive industries. This article mostly discusses these five papers, using

14

them as an exemplar only to highlight the importance and use of Artificial Neural Network (ANN) in Friction Stir Welding (FSW) process.

Tansel et al [9] used genetically optimized neural network systems (GONNS) to estimate the optimal operating condition of the friction stir welding (FSW) process. He introduced the genetically optimized neural network system (GONNS) by using Artificial Neural Network (ANN) and Genetic Algorithm (GA) together. He represented Friction Stir Welding (FSW) process in five artificial neural networks (ANN) as shown in the Figure 2. The genetically optimized neural network is shown in the Figure 3. Artificial Neural Network (ANN) is first trained by the genetically optimized neural network systems (GONNS) with experimental data. . It was observed that the inputs of the five ANNs were the same (tool rotation and welding feed rate). The estimation errors of the ANNs were better than average 0.5%. GA estimated the optimal FSW conditions to minimize or maximize one of the stir welding characteristics, while the others were kept at the desired ranges.

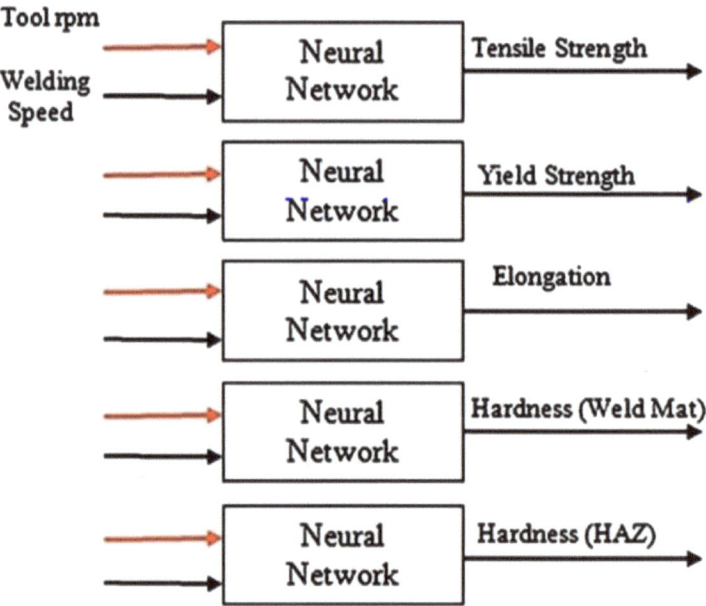

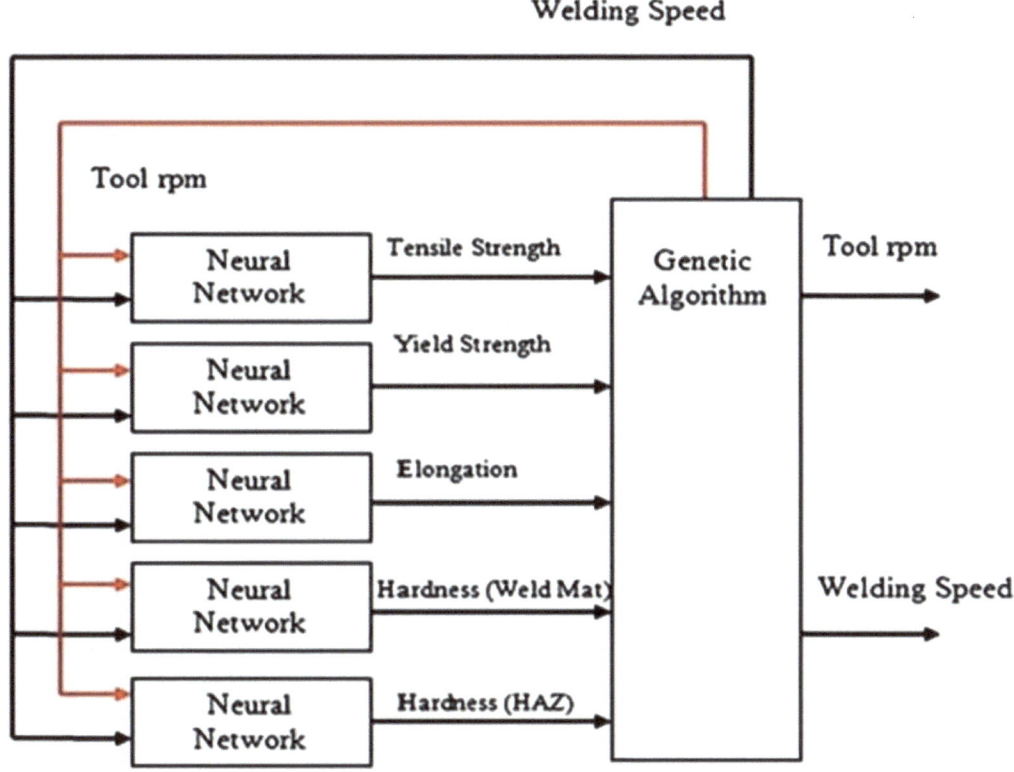

Dehabadi et al [10] used tow Artificial Neural Network (ANN) to study the effects of thread and conical shoulder of each pin profile on the micro hardness of welded zone of AA6061 plates as shown in the Figure 4.

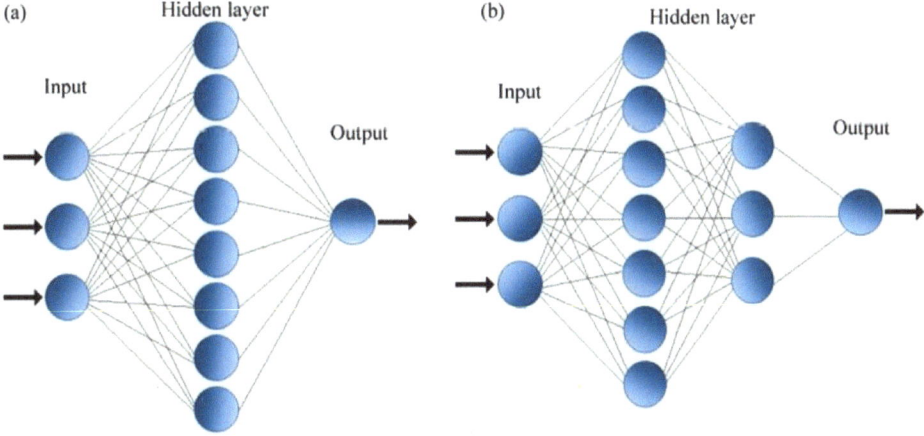

It was observed that the Mean absolute percentage error (MAPE) for train and test data sets did not exceed 5.4% and 7.48%, respectively. MSE values for both networks were less than 10, which indicated appropriate trained models as shown in the Figure 5.

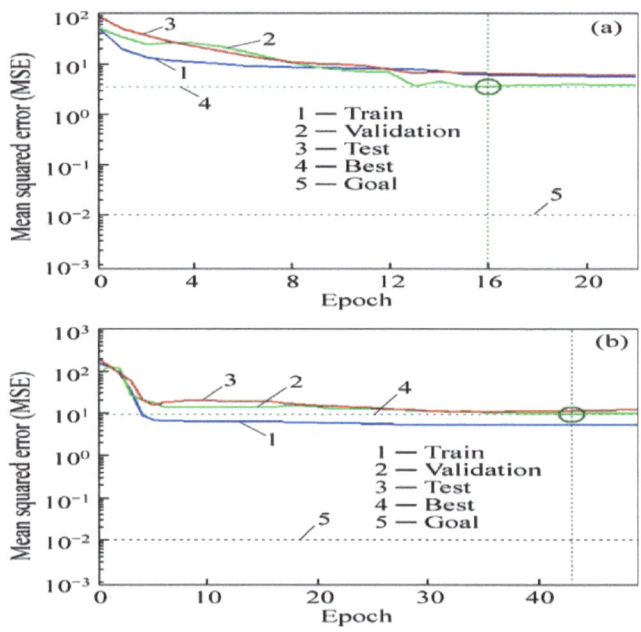

Shojaeefard et al [11] numerically modelled a different tool pin and shoulder diameter for a Friction Stir Welding (FSW) process. He used Feed-forward neural network with back-propagation algorithm to understand the correlation between tool dimensions and peak temperature, maximum strain, and HAZ area as shown in the Figure 6.

It can be observed that the neural network having an input layer with two neurons for each input factor (pin diameter, shoulder diameter) and an output layer with three neurons (maximum strain, maximum temperature, and HAZ area) was used. In order to get the best network architecture evaluation of several architectures were performed and trained using the experimental data. Based on this analysis, the optimal architecture was selected as 2–6–2 NN, and both activation functions in hidden layer and output layer were ''logsig.''

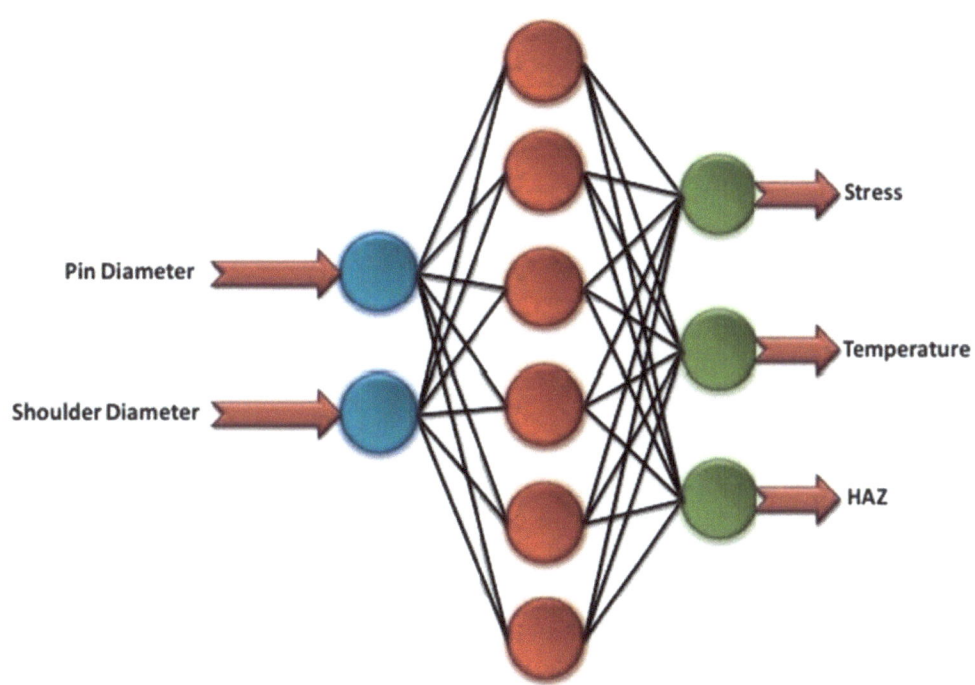

Fratini et al [12] in his research showed the capability of the AI technique in conjunction with the FE tool to predict the final microstructure of the Friction Stir Welded Joints. He designed the network architecture which was composed of five layers as shown in Figure 7. From the Figure 7 it is clearly seen that the neural network consisted of an input layer, three hidden layers and finally the output layer. Input layer was composed of four neurons which represented the local values of the equivalent plastic strain, the strain rate, temperature and the Zener Holloman parameter in a transverse section. The introduced three hidden layers have three, five and four neurons, respectively, and finally in the output layer one neuron is present corresponding to the output variable (D), namely the local value of the final average grain size. Each layer is fully connected to the next and according to the back propagation rule, the weights (wij) of the connections linking neurons belonging to two consecutive layers are adjusted in the learning stage with the aim to minimize the error between the desired output and the calculated one.

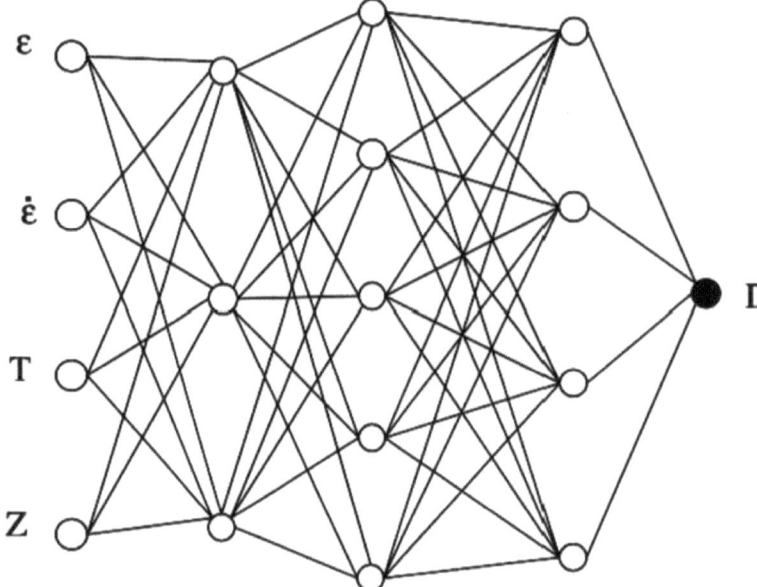

Jayaraman et al [13] pedicted the tensile Strength of Friction Stir Welded A356 Cast Aluminium Alloy by using Response Surface Methodology (RSM) and Artificial Neural Network (ANN). He used the topology architecture of feed-forward three-layered back propagation neural network as shown in the Figure 8. He noted that the performance of Artificial Neural Network (ANNs) is better than the other techniques, especially RSM when highly non-linear behaviour is the case. Also, this technique can build an efficient model using a small number of experiments; however the technique accuracy would be better when a larger number of experiments are used to develop a model.

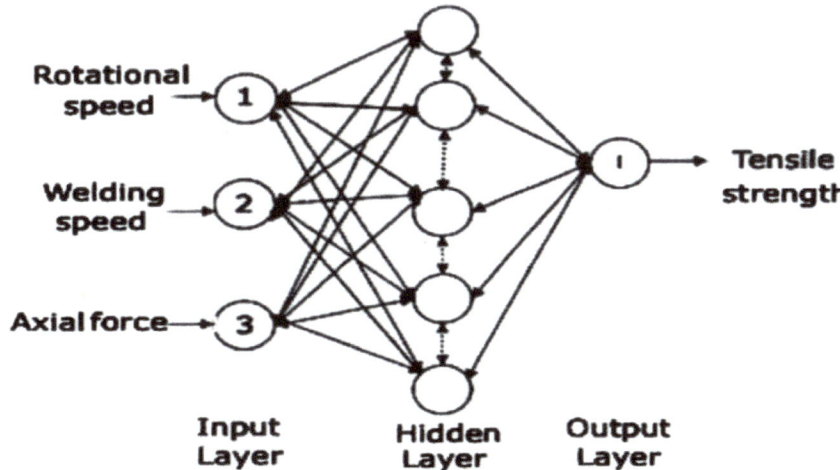

Maleki et al [14] used ANN as an efficient approach for modeling the mechanical properties of Friction Stir Welded 7075-T6 Aluminum alloy. ANN devloped was based on Back propogation algorithm. Rotational speed of tool, welding speed, axial force, shoulder diameter, pin diameter and tool hardness are regarded as inputs of the ANNs. Yield strength, tensile strength, notch-tensile strength and hardness of welding zone are gathered as outputs of neural networks as shown in the Figure 9.

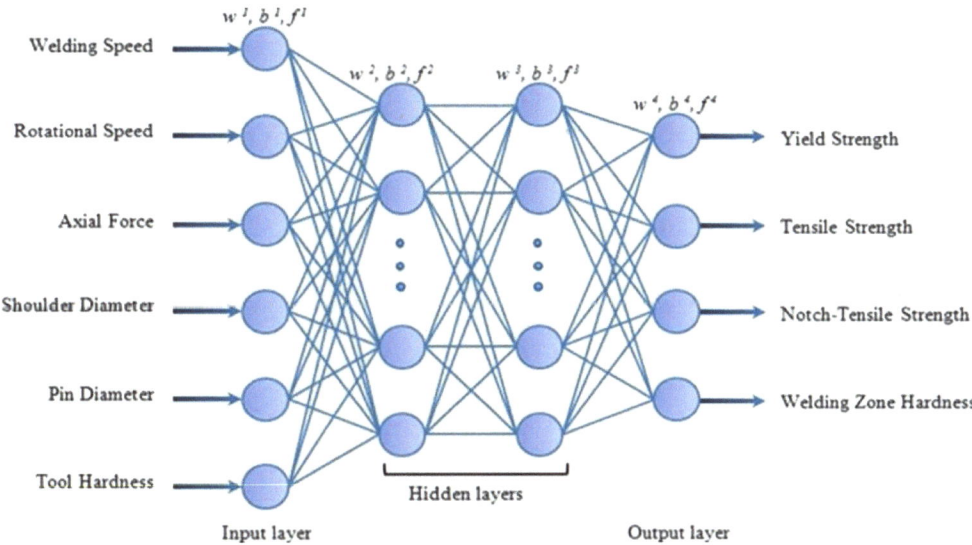

The least mean relative error (MRE) was obtained for the hardness of welding zone, yield strength, tensile strength and notch-tensile strength.

Khoursid et al [15] used the topology architecture of feed-forward three-layered back propagation neural network as illustrated in Figure 10 below for predicting the ultimate tensile strength, percentage of elongation and hardness of 6061 aluminum alloy.

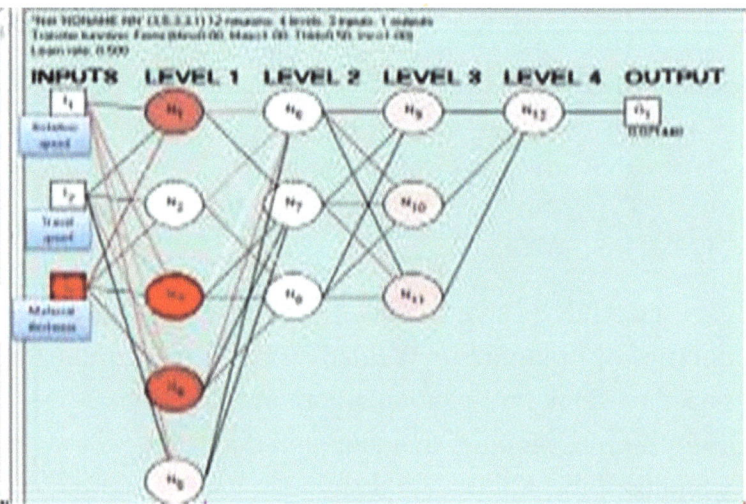

Equation is calculated as On = F(ΣIk * Wkn).On is the neuron's output, n is the number of the neuron, Ik are the neurons inputs, k is the number of inputs, Wkn are the neurons weights. F is the Fermi function 1/(1+Exp(-4*(x-0.5))).

Software (pythia) was used for training the network model for tensile strength, the percentage of elongation and hardness prediction. The neural network described in this paper, after successful training, will be used to predict the tensile strength of friction stir welded joints of 6061 aluminum alloy within the trained range. The results obtained after training and testing On artificial neural networks are shown in the Fig.(11-13).

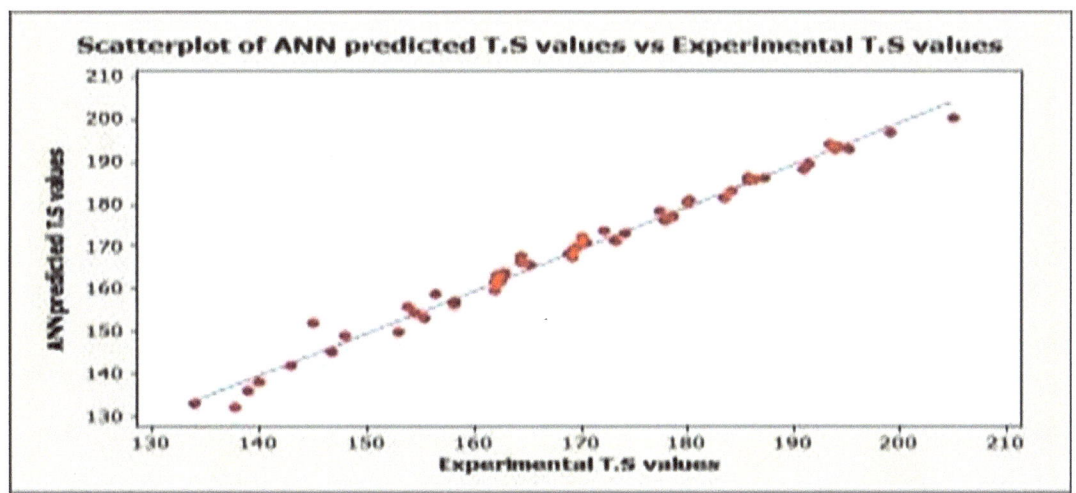

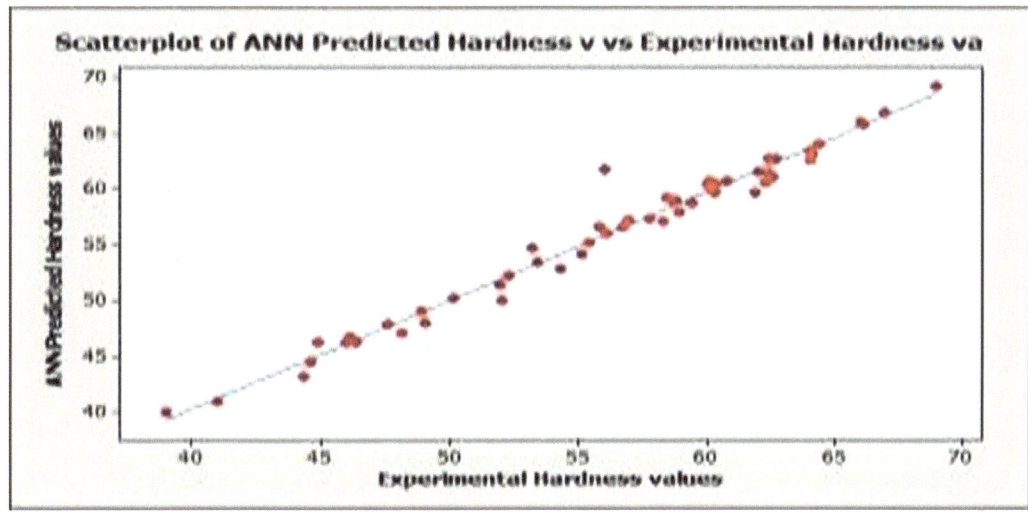

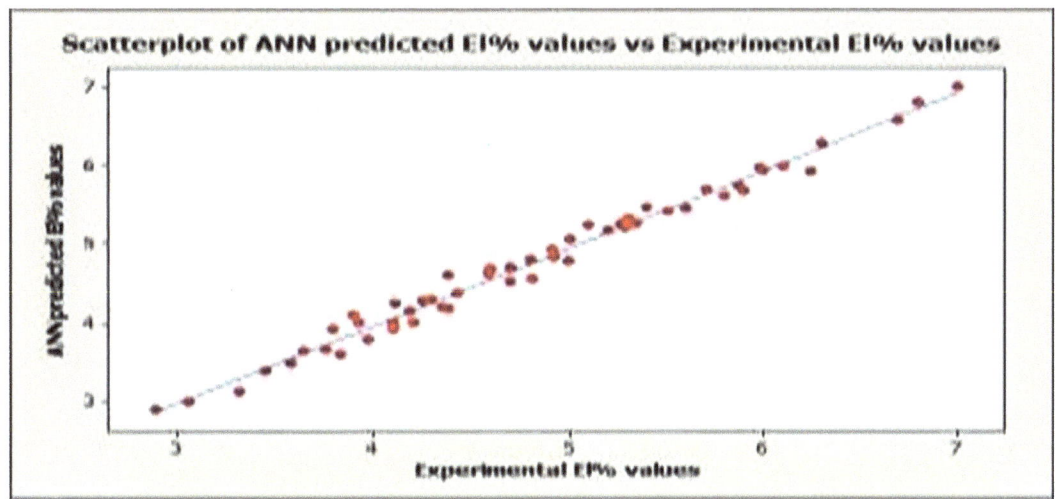

The ANN model proved to be successful in terms of agreement with experimental results ratio 96.5%.

H. Okuyucu et al. [16] developed an artificial neural network (ANN) model for the analysis and simulation of the correlation between the friction stir welding (FSW) parameters of aluminium (Al) plates and mechanical properties. The input parameters of the model consist of weld speed and tool rotation speed (TRS). The outputs of the ANN model include property parameters namely: tensile strength, yield strength, elongation, hardness of weld metal and hardness of heat effected zone (HAZ). Good performance of the ANN model was achieved. The model can be used to calculate mechanical properties of welded Al plates as functions of weld & tool rotation speeds. The combined influence of weld speed and TRS on the mechanical properties of welded Al plates was simulated. A comparison was made between measured and calculated data. The calculated results were in good agreement with measured data. The aim of the paper was to show the possibility of the use of neural networks for the calculation of the mechanical properties of welded Al plates using FSW method. Results showed that, the networks can be used as an alternative in these systems.

L Fratini and G Buffa [17] studied the continuous dynamic re-crystallisation phenomena occurring in the FSW of Al alloys. A good agreement with the experimental results was obtained using the ANN model. In regard to ANNs, it noted that ANNs perform better than the other techniques, especially RSM when highly non-linear behaviour is the case. Also, this technique can build an efficient model using a small number of experiments; however the technique accuracy would be better when a larger number of experiments are used to develop a model.

References

1. Maind, S.B. and Wankar, P., 2014. Research paper on basic of artificial neural network. International Journal on Recent and Innovation Trends in Computing and Communication, 2(1), pp.96-100.

2. Sha, W. and Edwards, K.L., 2007. The use of artificial neural networks in materials science based research. Materials & design, 28(6), pp.1747-1752.

3. Sato, Y.S., Kokawa, H., Enomoto, M. and Jogan, S., 1999. Microstructural evolution of 6063 aluminum during friction-stir welding. Metallurgical and Materials Transactions A, 30(9), pp.2429-2437.

4. Lienert, T.J., Stellwag Jr, W.L., Grimmett, B.B. and Warke, R.W., 2003. Friction stir welding studies on mild steel. WELDING JOURNAL-NEW YORK-, 82(1), pp.1-S.

5. Mishra, Akshansh, et al. "Mechanical and Microstructure properties analysis of Friction Stir Welded Similar and Dissimilar Mg alloy joints." (2018).

6. Mishra, Rajiv S., and Z. Y. Ma. "Friction stir welding and processing." Materials science and engineering: R: reports 50.1-2 (2005): 1-78.

7. Ouyang, Jiahu, Eswar Yarrapareddy, and Radovan Kovacevic. "Microstructural evolution in the friction stir welded 6061 aluminum alloy (T6-temper condition) to copper." Journal of Materials Processing Technology 172.1 (2006): 110-122.

8. Dressler, Ulrike, Gerhard Biallas, and Ulises Alfaro Mercado. "Friction stir welding of titanium alloy TiAl6V4 to aluminium alloy AA2024-T3." Materials Science and Engineering: A 526.1-2 (2009): 113-117.

9. Tansel, Ibrahim N., et al. "Optimizations of friction stir welding of aluminum alloy by using genetically optimized neural network." The International Journal of Advanced Manufacturing Technology 48.1-4 (2010): 95-101.

10. Dehabadi, V.M., Ghorbanpour, S. & Azimi, G. J. Cent. South Univ. (2016) 23: 2146. https://doi.org/10.1007/s11771-016-3271-1

11. Shojaeefard, Mohammad Hasan, et al. "Investigation of friction stir welding tool parameters using FEM and neural network." Proceedings of the Institution

of Mechanical Engineers, Part L: Journal of Materials: Design and Applications 229.3 (2015): 209-217.

12. Fratini, Livan, Gianluca Buffa, and Dina Palmeri. "Using a neural network for predicting the average grain size in friction stir welding processes." Computers & Structures 87, no. 17-18 (2009): 1166-1174.

13. Jayaraman, M., et al. "Prediction of tensile strength of friction stir welded A356 cast aluminium alloy using response surface methodology and artificial neural network." Journal for Manufacturing Science and Production 9.1-2 (2008): 45-60.

14. E Maleki 2015 IOP Conf. Ser.: Mater. Sci. Eng. 103 012034

15. A.M. Khourshid , Ahmed. M. El-Kassas , H. M. Hindawy and I. Sabry, MECHANICAL PROPERTIES OF FRICTION STIR WELDED ALUMINIUM ALLOY PIPES, European Journal of Mechanical Engineering Research, Vol.4, No.1, pp.65-78, April 2017

16. Hasan Okuyucu a, Adem Kurt a, Erol Arcaklioglu -Artificial neural network application to the friction stir welding of aluminum plates- Materials and Design 28 (2007) 78–84

17. .L Fratini and G Buffa- Continuous dynamic recrystallization phenomena modelling in friction stir welding Proceedings of the Institution of Mechanical Engineers; May 2007; 221, B5; ProQuest Science Journals pg. 857.

www.ingramcontent.com/pod-product-compliance
Lightning Source LLC
Chambersburg PA
CBHW041308180526
45172CB00003B/1013